TABLE OF CONTENTS

CHAPTER   PAGE

ABSTRACT ii

    I   INTRODUCTION 1

    II   BACKGROUND 3

    III   ASSUMPTIONS AND ANALYSIS OF KEY FACTORS 9

    IV   RECOMMENDATIONS AND CONCLUSION 18

NOTES 24

BIBLIOGRAPHY 26

CHAPTER I

INTRODUCTION

The polar icecap which covers the Arctic Ocean is melting. It is a well-known, scientific fact. Global warming is the generally accepted cause for the big meltdown that is going on at the top the world. Along with the retreating ice coverage, the ice is getting thinner. One of the very likely results of this environmental phenomenon is the potential year-round opening of Arctic Sea routes, which would connect the Atlantic and Pacific Oceans and provide shorter distances for inter-theater movement of naval forces. The two existing sea routes are the Northeast Passage, or the Northern Sea Route, which follows the Russian coastline, and the fabled Northwest Passage, which runs through Canadian waters. Some predictions suggest that unassisted shipping (ships transiting without icebreaker escort) could transit these passages in as soon as five to ten years during the warmer summer months, and transit on a year-round basis in fifty years.

The strategic and operational implications for U.S. national security strategy regarding blue waters in the Arctic region are significant. Most obvious is that the space between the U.S. and Russia could be far more accessible for both surface and subsurface naval forces. U.S. interests in new oil and mineral exploitation along with new Arctic freedom of navigation issues could likely mandate a need for increased U.S. presence in the region. At the operational warfighting level, most significant would be quicker maritime options for theater-to-theater strategic deployments and force projection for a supported combatant commander's major operation or campaign. Future use of the Arctic Sea routes would directly support U.S. national security strategy, in force projection and strategic mobility for response to threats and crises.

The purpose of this paper is to analyze key elements and provide recommendations in consideration of the Arctic Sea routes for future operational use. Moreover, this paper explores issues which are on the horizon -- both in the near term and several decades into the future -- which are relevant in the concept of joint force projection through the Arctic region. In considering these routes for force projection and strategic agility, specific concerns and issues identified and discussed include assumptions regarding key elements, the application of the Law of the Sea, and operational factors and functions, as applicable.

CHAPTER II

BACKGROUND

Global Warming and the Arctic

During this and the last century, researchers have found substantial evidence that the Arctic region is, in fact, warming. Alaskan soil samples have indicated a substantial temperature increase. Even the deep layers of the Arctic Ocean are warming.[1] In the past twenty years, average annual temperatures in the Arctic region have risen some 7 to 9 degrees Fahrenheit. Along with the rise in temperature, the Arctic Sea ice is some 40 % thinner and covers about six percent less than it did in the early 1980s. The primary cause of this meltdown at the North Pole is, according to many scientists, due to the world's increased burning of fossil fuels, such as gasoline and coal, which has overloaded the atmosphere with carbon dioxide and other greenhouse gases.[2] That "greenhouse effect" has trapped the sun's heat in the earth's atmosphere, causing surface and air temperatures to rise. Globally, the carbon dioxide level which exists today is estimated to be some 30 % higher than it was before the industrial age.[3] All indications are that global warming will continue, and that the Arctic ice will continue to melt. As the polar icecap shrinks, the two semi-circular sea routes along its outer edge will increasingly become more ice-free. Were the icepack to melt altogether, an Atlantic to Pacific sea route directly through the North Pole would exist.

The Arctic Sea Routes

<u>Northeast Passage</u>. The Northeast Passage, more commonly known as the Northern Sea Route, provides a link between the Atlantic and Pacific via a total of five Arctic seas which exist primarily along the Russian coast. With a starting point at Murmansk, Russia (east of Finland, and east of the Norwegian Sea), the main lane of the route runs through the Barents Sea, passing north of Novaya Zemla, and proceeds across the Kara Sea into the Vilkitsky Straits and then into the Laptev Sea. From there, the route leads into the East Siberian Sea and then on through the Bering Strait into the Bering Sea. Of the five seas which collectively comprise the Northern Sea Route, four have average depths of two hundred meters or less. On both ends of the route, the Bering and Barents Seas are virtually ice-free. In the Kara Sea, the icepack reaches to Severnaya Zemlya, often requiring passage through the Vilkitsky Straits, which is 11 miles wide at its narrowest point.[4]

The Northern Sea Route was first used to sail from the Atlantic to the Pacific in 1878. While it has never been used as a dedicated international route due to the presence of ice for much of the year, Russia has been very successful in using the route for transfer of cargo along the Russian coast and elsewhere as the ice permits. Moreover, Russia has over the years built a very capable fleet of icebreakers and also established a network of fixed and mobile Arctic stations.[5] As such, Russia is ahead of the U.S. in the Arctic region, in terms of presence and capabilities.

  <u>The Northwest Passage</u>. Along with the latest developments regarding the environmental impacts of global warming and the big meltdown in the Arctic region, much has recently been written about the potential for the Northwest Passage to connect the Atlantic and Pacific oceans. For over four hundred years, it has, for the most part, remained an elusive quest to explorers and navigators. The passage was discovered by British explorer Sir John Franklin in 1845 (his expedition ended in cannibalism). Not until 1906 did Norwegian explorer Roald Amundsen (also the first to the South Pole) complete a three-year journey through the passage. Ever since, the passage has been a topic of commercial interest. The distance for shipping between Europe and the Far East could be cut from the current Panama Canal route of 12,600 miles to a far more attractive 7,900 miles.[6] That's a savings of from 10 to 15 days of straight transit time, depending on the speed made good over the ground. For larger ships unable to transit through the Panama or Suez Canals, the Europe to Far East distance saved would be 6,770 miles when compared to the route around Africa -- a savings of between 15 to 20 days of transit time.

  Four transit lanes have been developed in the Northwest Passage so far, crossing from the Atlantic to the Pacific, and vice versa. The lane which has the most likelihood of being used for international navigation is the route commonly referred to as the Northwest Passage. That route commences in the east at Lancaster Sound and runs west through Barrow Strait and Viscount Melville Sound. From there, the route runs southwest through Prince of Wales Strait between Banks and Victoria Islands, and then west again along the north coast of Canada and Alaska, then to and through the Bering Strait.[7]

The width of the Northwest Passage ranges from more than 75 miles to five miles at the narrowest point. Currently, the navigation season lasts about four months, with the help of Canadian icebreakers. Otherwise, depending on wind and ice movements from the Beufort Sea, the route has been closed for the remainder of the year.[8]

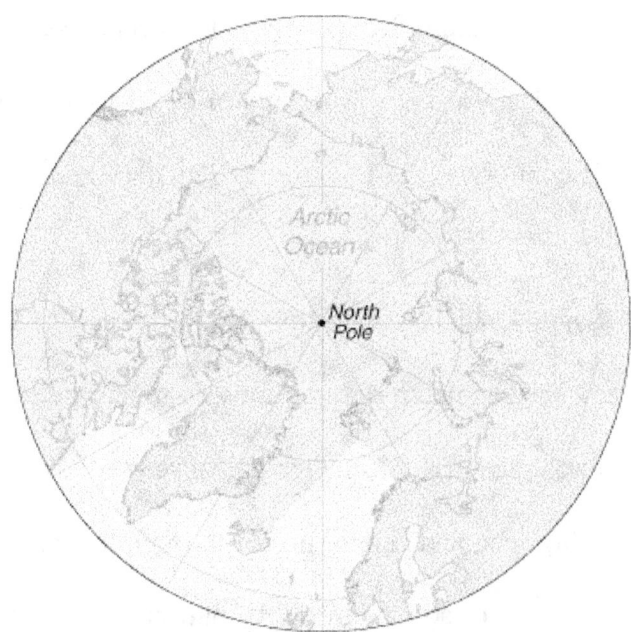

Figure II-2. Arctic Passages

Applicability of the Law of the Sea to the Arctic Sea Routes

General. There are several, ongoing legal issues between the United States, Canada, and Russia involving the Law of the Sea and its applicability to both the Northeast Passage and the Northwest Passage. Specific issues involve territorial seas and innocent passage, transit passage, freedom of navigation in international waters, and international strait determinations. Of interest is that the United States' position regarding freedom of navigation has been demonstrated several times by recent transits of Coast Guard icebreakers through both passages, both with and without consent.

The Northeast Passage. The right of innocent passage applies to all ships transiting the Northeast Passage. Although the Vilkitsky Straits clearly are well within Russian territorial seas, the right of innocent passage still applies. Since the Vilkitsky Straits have not been determined to be international straits used for international navigation, Russia retains the right to impose coastal laws regulating innocent passage, as well as temporary suspension of passage for national security reasons. Russia has additionally legislated that passing warships in peacetime (including U.S. Coast Guard icebreakers) are subject to complete discretion and control in transiting Russian territorial seas. This view regarding warship innocent passage is contrary to the Law of the Sea, and has doubtful validity in international law.[9]

The Northwest Passage. Canada has, for the long term, asserted sovereignty over the passage. Moreover, Canada has claimed a right to draw straight baselines around the perimeter of the Canadian archipelago, limit the territorial seas and, further, exercise complete control of which ships can and cannot transit through the passage. Canada's apparent concern is a combination of sovereign rights, national territory, and protection of the fragile Arctic environment. The United States has, for the long term, contended that the passage is in international waters, and has pressed for freedom of navigation and designation as an international strait.[10] The current legal summary status is that the right of innocent passage applies to all ships transiting the passage, in accordance with the United Nations Convention on the Law of the Sea (UNCLOS). Canada could suspend passage of foreign ships if it is essential to national security, which has been determined to include anti-pollution standards. In peacetime, warships may be expelled for non-compliance with Canada's security regulations.[11]

U.S. National Security Interests in the Arctic Region. Starting with the under-ice transit of the Arctic Ocean by the U.S.S. Nautilus in 1957, the Arctic Region became increasingly important as a deployment area for both American and Russian strategic ballistic missile submarines. The region was considered highly valuable because the ice provided cover from air and satellite detection, and the ice pack interfered with underwater detection technology--sonar, sound, and magnetic. The general idea was to strategically deploy SSBNs in the Arctic, where they were at the ready to surface in thin ice or openings in the ice, and fire ballistic missiles at the enemy. As a result of this mutual underwater threat, the U.S. established an elaborate system of underwater listening systems in the Norwegian Sea, the North Atlantic, and the North Pacific. Prior to the collapse of the Soviet Union, the air threat to the U.S. from the Arctic region was likewise considered high. As such, the North American Air Defense Command was established, consisting of some thirty radar sights, and headquartered at Colorado Springs.[12]

With the former Soviet Union no longer considered a major U.S. national security threat, interest in the Arctic has given way to other priorities. Over the years, several agencies and groups were established by legislation to coordinate and implement U.S. policy regarding activities and programs in the Arctic Region. These agencies and groups include the Interagency Arctic Policy Group (IAPG), which was followed by the Interagency Arctic Research Policy Committee (IARPC).[13] In 1996, the eight Arctic governments (U.S., Canada, Russia, Denmark/Greenland, Finland, Iceland, Sweden, and Norway) established the Arctic Council to address issues of mutual concern. The United States' ongoing, post-Cold War security interest in the Arctic region has been the preservation of freedom of the seas and the superjacent airspace.

CHAPTER III

ASSUMPTIONS AND ANALYSIS OF KEY FACTORS

The focus of this section is the potential operational application of Arctic Sea routes to force projection and strategic agility. As defined in the 1997 National Military Strategy, force (or power) projection is "the ability to rapidly and effectively deploy and sustain U.S. military power in and from multiple, dispersed locations until conflict resolution."[14] Strategic (or global) agility is defined as the "timely concentration, employment and sustainment of U.S. military power anywhere, at our own initiative, and at a speed and tempo that our adversaries cannot match...it allows us to conduct multiple missions, across the full range of military operations, in geographically separated regions of the world."[15]

Due to the operational factors of time and distance, the potential operational advantages of the Arctic Sea routes are apparent in a major operation or campaign, in which the supported CINC requires both immediate and potentially long-term, sustained forces and support in large amounts, and over large distances. Force projection through the Arctic Sea requires analysis of certain elements of operational design, with a focus on the operational factors of space and time and the operational functions of movement and deception. Other areas to consider include movement through the Arctic region as compared to movement through the Panama Canal, the geographic and functional CINCs' interests in the Arctic region, and the potential expansion of functional and operational roles of the Alaskan Command, during inter-theater operations.

In a future major regional conflict (MRC) scenario, whether in Southeast Asia, the Middle East, or other regions of the world, the level of inter-theater military might and sustained support would likely surpass the levels which were required for the Gulf War. Moreover, the forces and support required for a prolonged campaign can easily be compared to quantities and levels used in Vietnam and Korea. In a future MRC, U.S. force projection and strategic mobility through the Arctic Sea could be a significant factor in decisive action. The Arctic Sea routes would merit serious consideration in the deliberate planning process, when an alternative to the Panama Canal may be needed, and when movement of naval forces and sealift requirements for sustained logistics are at high levels.

Assumptions. In considering future use of the Arctic Sea routes for force projection and strategic mobility, the following assumptions are made:

1. Due to the effects of global warming, the Arctic Sea ice will continue to melt, opening the sea routes for year-round Atlantic to Pacific transit, and potentially opening up the entire Arctic Sea.

2. That the principle element of responding to threats and crises will continue to be part of the overall national security strategy for enhancing U.S. security at home and abroad. Moreover, that strategic agility will continue to be critical in the ability to augment forces already forward deployed in global regions, with additional forces for international and domestic crisis response.[16]

3. That future force projection and strategic agility through the Arctic region would be in support of U.S. national security strategy.

4. That for force projection and sustained logistics in a regional theater, more than 90 % of the lift requirements will be accomplished by sealift.[17] Moreover, that sealift assets and capability will not substantially change in the next twenty to thirty years, if not longer.

5. That, due to across-the-board military reductions and the associated reduction in global forward presence, a MRC scenario would require substantial force projection. Further, that force projection will continue to be joint in nature, involving movement of land, air, and sea components, by sealift and airlift. Additionally, that force projection will involve coalition forces, when and where applicable.

6. That future regional conflicts in which the U.S would likely be involved, and which would require inter-theater force projection over significant distances, include the Korean Peninsula, Taiwan, and the Middle East. Other areas include the Balkans, Africa, India and Pakistan, and Southwest Asia.

7. That some capability for force projection through the Arctic Sea routes would exist, in response to the above potential conflicts, in approximately five to ten years. The passages will likely be ice free or nearly so in the summer months, and icebreaker-assisted force projection could be available for the remainder of the year.

8. That China is potentially a rising peer competitor, with a national strategy aimed at regional hegemony, and an associated military strategy aimed at both defense of claimed sovereign seas and territories and eventual force projection.

9. That by the time China is capable of force projection, the Arctic Sea routes could be available for projection of both U.S. and Chinese forces.

10. That an alliance or coalition between China and Russia would obviously increase the risks associated with force projection through the Arctic in terms of force protection. Other considerations regarding such an alliance or coalition would

include an increased U.S. defensive posture in the Arctic, and unavailability of the Northern Sea Route for force projection.

<u>Factors Space and Time</u>. In considering the Arctic Sea routes for U.S. force projection--for example, in a war scenario with China in which Taiwan has been invaded-- the factor of space is significant, specific to the distance required to project the power needed. As such, the distance in moving forces from U.S. departure points to Japan as a probable base of operations is closely related to the factor of time. The larger the distance is, the more complicated and longer it is to project power.[18] Naval forces departing from the U.S. East Coast and from the Mediterranean would have considerable distances to transit in order to arrival in the theater of operations.

Other important considerations regarding factor space in the Arctic Sea lines of communication are the hydrography and oceanography of the Arctic region. Analyzing sea conditions, sea ice, thermal structure, bottom depth, and so on are fundamental in the operational planning of maritime asset movement. Other factors to consider include climate, weather, and channels and passages which, especially in the Arctic, could have a direct bearing on operational sequencing to the planned battlespace. A potential advantage of the Arctic is that, while it is a harsh climate, it is not susceptible to hurricanes and typhoons, such as the North Atlantic and Pacific regions are. Lastly, a general view of factor space and the Arctic region is that, due to human and state interests and resulting activities and conflicts which did not exist before because of inaccessibility, the Arctic region inherently becomes new potential battlespace.

Factor time is the most important operational consideration in analyzing the Arctic Sea routes for inter-theater force projection. For offensive operations, more time is needed to start the deployment and build up forces in a concentration area, followed by maneuver against the enemy. Factor time is of critical importance for the offense -- the less amount of time it takes to move forces for concentration and maneuver, the less time the defender has to prepare for a defensive strategy. Factor time becomes even more critical the larger the distance is for the movement and deployment of forces and supporting logistics.[19] In a Southeast Asia MRC scenario, the distance required for force movement by sea in support of U.S. Pacific Command (CINCPAC) is considerable. The following table provides approximate distances and estimated times required in the movement of supporting forces from Norfolk and the Mediterranean to Japan, the base of operations.

Strategic Deployment Distance Table

Departure Point	BOO*	Route	Estimated Distance	Ave SOG**	Transit Time
Norfolk	Japan	Arctic	8,500 NM	24	14.8 days
Norfolk	Japan	Panama	10,000 NM	24	17.2 days
Norfolk	Japan	S Amer.	16,500 NM	24	28.9 days
Norfolk	Japan	Med	13,000 NM	24	22.9 days
Med	Japan	Arctic	9,000 NM	24	15.6 days
Med	Japan	Panama	13,500 NM	24	23.3 days
Med	Japan	Suez	9,000 NM	24	15.6 days
Med	Japan	S Africa	18,000 NM	24	31.3 days

* BOO - base of operations
** SOG - speed over ground

Using the above table, as an example, moving a CVN battle group from the Mediterranean to Japan, via the South Africa route (Suez Canal not open and CVN too large to transit the Panama Canal) would take 31.3 days straight steaming time. Moving a Maritime Action Group from Naval Station Norfolk to Japan via the Panama Canal would take 17.2 days. The aggregate transit time would be 48 days, as compared to 30 days for the same movement through the (ice free) Arctic Sea. Moving large numbers of supporting forces over great distances in the least amount of time results in more opportunities for freedom of action by the operational commander.

<u>Synthesis of Operational Factors</u>. Factors space and time are directly related to strategic agility throughout the world and force projection provided to supported CINCs in regional conflicts. The relationship of space and time specific to the Arctic Sea routes results in the trade of distance, in terms of sea space required for strategic deployments, for time, in terms of time it takes for surface and subsurface assets to transit from departure points to the operational theater. In essence, the Arctic Sea routes would cut time in the movement of maritime assets from virtually any place in the world from a departure point on or above the mid-latitudes, by transiting over the top of the world.

The relationship between time and distance in force projection should not considerably change in future years. In considering the potential future use of technological advances in shipping--e.g., ultra-fast sealift ships--which would reduce time by increasing speed over distances, the effect of transiting through he Arctic Sea routes would compound the factor time. That is, less time by greater speed and less distance would equate to an overall greater reduction of time.

<u>Operational Design - Deployment and Deception</u>. Regarding operational deployment, the Arctic Sea routes would potentially provide an alternate route from the U.S. East Coast and Europe, other than the Panama Canal. Perhaps even more significant is that the design of operational deployment could be fundamentally changed in an Atlantic-to-Pacific movement of maritime assets in that forces could be moved to the operational theater from two directions--from the mid-latitudes (Panama Canal), and from the top (Arctic Sea).

During the mid 20th century, U.S. Navy surface ships were designed and built to fit through the Panama Canal. A modern-day reference to former Navy shipbuilding was only recently evidenced by the movement of U.S.S. Iowa from Newport to San Francisco via the Panama Canal--with four feet to spare on both sides of the ship while in the locks. Today's CVNs were designed and built too big to get through the Panama Canal, primarily because modern carrier aircraft need longer flight decks for launching and recovering. Additionally, the underpinnings of forward presence throughout the regions of the world eliminated much of the supposed need to move forces from theater to theater. Expanded U.S. forward presence force requirements in today's security environment--meaning that forces are spread thin globally--almost guarantee the need to move significant amounts of forces to the supported CINC'S area of responsibility. Moreover, the likelihood of sustained 7 x 24 carrier air operations requirements in a major operation or campaign could in itself drive inter-theater carrier battle group movements.

The Arctic Sea routes could provide the opportunity for operational deception, as well. With media coverage of a movement through the Panama Canal, forces could be moved through the Arctic with little attention. Specific to submarine forces which transit through the Panama Canal on the surface, the Arctic provides the opportunity to transit below the surface, undetected.

Arctic Sea Routes and CINC Interests. While CINCPAC has current strategic interests in the Arctic, both CINCPAC and U.S. Joint Forces Command (CINCJFCOM) will likely have increasing interests in the Arctic region, due to both the potential of force projection and strategic agility through the Arctic, and also due to the general increase of commercial activity and U.S. interests in the region. The current geographic CINC boundaries divide the Arctic region in two, half to Pacific Command and half to Joint Forces Command. As such, the Northwest and Northern Sea Routes are likewise divided. U.S. Southern Command (CINCSOUTH) would likely have a secondary interest in the Arctic region, which could provide an alternate--or perhaps primary--route instead of the Panama Canal. U.S. European Command (CINCEUR) would clearly have an interest specific to the Arctic Sea routes, in view of the relationship between NATO and the Arctic Council membership (particularly Canada). For all operational, supporting CINCs, the general interest in the Arctic Sea routes is the phasing and sequencing of forces and supplies to the supported CINC. For the functional CINCs, U.S. Transportation Command (TRANSCOM) would have substantial interest in the Arctic due to sealift mission requirements for sustained logistics.

The Alaskan Command. The Alaskan Command, under PACOM, consists of U.S. Air Force (air), Army (ground), and Coast Guard (naval) forces. Air forces at Elmendorf AFB, near Anchorage, and at Eielson AFB, near Fairbanks, provide air defense and air superiority in Alaska, in addition to providing ready air support to PACOM and the Unified Commanders. U.S. Army Alaska (former 6th Infantry Division - Light), headquartered at Fort Richardson, supports contingency operations in PACOM. The Seventeenth Coast Guard District Commander, Anchorage, serves as U.S. Naval Forces Alaska. The Coast Guard maintains air stations at Kodiak and Sitka, and homeports major cutters, patrol boats and buoy tenders throughout Alaska.

Were the Arctic Sea routes to be used in conjunction with U.S. strategic agility and operational force projection, the Alaskan Command's breadth of functional responsibilities would widen considerably. Additional or increased functions would likely include increased or even sustained icebreaking capability and force protection. Logistically, a support base would be needed. Nome or Adak, Alaska would be logical candidates.

While the big meltdown in the Arctic has the potential to offer alternatives for strategic mobility and force projection for the U.S., it offers the same potential to the enemy. The additional functions along with the potential for sustained Arctic defense could result in a larger sub-unified command, comparable, for example, to U.S. Forces Korea. Of note is that the Alaskan Command's strategic responsibilities would already likely expand if and when the National Missile Defense System is deployed.

<u>Status of the Panama Canal</u>. In 1999, the U.S. ended 96 years of military presence and control of the Panama Canal, and transferred ownership of the canal to the Republic of Panama. Under the existing treaty, the U.S. maintains the right to intervene if the neutrality or the security of the canal is threatened. Moreover, under the treaty, U.S. warships are guaranteed "expedited" and "head-of-the-line" passage during war.[20]

Of potential concern is the ability of the Panamanian government, with its history of instability, to maintain both the neutrality and security of the Canal. As the Canal continues to age, maintenance and upkeep--or lack thereof--are additional, potential major concerns. Of further interest is that a Chinese company was granted control of the canal's two U.S.-built ports, on the Pacific and Atlantic sides. For U.S. force projection and sustained logistics primarily by sealift, the reliance on the Panama Canal is significant. U.S. and coalition military presence in Panama would likely be a requirement in a campaign or major operation involving inter-theater force projection through the Panama Canal. As such, the Arctic Sea routes may provide not only a viable, but potentially a better, alternative in future years.

CHAPTER IV

RECOMMENDATIONS AND CONCLUSION

In consideration of assumptions and analysis of key factors regarding potential future use of the Arctic Sea routes in support of U.S. national security strategy, it should at least be apparent that the idea needs much further study. My overall recommendation is to further assess the impact of global warming on the Arctic Sea ice and to develop a near-term course of action which would lay the foundation for future use of the Arctic Sea routes. Identified also is the foundation for a more long-term course of action, assuming that the Arctic Sea routes will become available for force projection and strategic agility in future years.

Near -Term Course of Action (5 to 10 Years)

Office of Naval Research and U.S. Arctic Research Commission. The Oceanographer of the Navy (N096), Office of Naval Research, National/Naval Ice Center, and the Arctic Research Commission should continue to monitor the effects of global warming in the Arctic, and provide ongoing predictions regarding near-future sea route availability and, in the long term, complete melting of the icepack.

What was to be a first recommendation has just recently been completed. The Oceanographer of the Navy held a symposium in April 2001 on the topic of naval operations in an iceless Arctic. Objectives of the symposium included: "(1) To begin study and analysis of strategic and policy issues which could elicit a military response due to the Arctic being sea ice free during a portion of the year; (2) Identify potential requirements for future naval operations given the projected retreat of the Arctic ice cap, and examine potential impacts and effects on such operations; (3) Identify baseline capabilities for operating in the altered Arctic environment; and (4) Establish the criteria and key elements for a continuum of heightened awareness and participation in examining operations in the altered Arctic environment."[21]

Law of the Sea Issues. The fundamental point regarding the Law of the Sea is that its conventions will clearly apply in an ice-free Arctic Sea, just like it does in other oceans and seas of the world. Of note is that a comparison to the treaties and conventions as applied to Antarctica is not valid, as Antarctica is a continental land mass surrounded by vast amounts of sea ice.

The U. S. State Department should continue to press for freedom of navigation in the Arctic Sea routes. Specifically, the U.S. should resolve the legal status of the Northwest Passage and Northern Sea Route. Important to force projection through these passages is the international strait legal determination. International strait status would permit the unimpeded surface and subsurface transit of U.S. forces, as opposed to potential control of warships desiring innocent passage, by coastal states in their territorial seas.

CINC Interests. As previously discussed, the U.S. Alaskan Command's responsibilities may increase with general, increased activity in the Arctic region. Further, it may be prudent for reconsideration of geographic boundaries of the Arctic region, currently divided between CINCPAC and CINCJFCOM. CINCJFCOM's current relationship with Arctic Council states combined with increased interests and issues in the Arctic might suggest expanding Joint Forces Command's geographic boundaries to include more of the region.

Future Course of Action (10 to 50 Years)

U.S. Coast Guard Forces and Missions. General, increased commercial and military activity in the Arctic region, along with potential future force projection through the Arctic Sea routes, may require increased Coast Guard presence in future years. Increased missions may include icebreaking, iceberg detection and tracking, maritime search and rescue, aids to navigation design and maintenance, and convoy escort. Planning factors for a future Coast Guard course of action would include determining adequacy of the icebreaker fleet (currently three are capable of polar operations), potential icebreaker weapons systems, and a general coastal and deepwater fleet capability to operate in ice conditions.

The Coast Guard plans to replace its aging major cutter fleet in this and the next decade with the service's Deepwater System Acquisition program. This new fleet will

Figure IV-1, USCGC Polar Sea

probably have a total life cycle of anywhere from thirty to fifty years, based on life cycles of previous cutter acquisitions. At the same time, the service's missions in the Arctic region will likely increase, particularly toward the end of the new fleet's life cycle. As such, the

Deepwater project "after next" should plan for Arctic operations, with matching force capabilities in extreme northern regions.

Additionally, the Seventeenth Coast Guard District's functional mission area of responsibility will expand in the Arctic regions, due to U.S. interests in shipping, oil and mineral exploitation, and fisheries enforcement. General maritime activity in the Arctic could drive the organizational and functional establishment of a Coast Guard Arctic Area, as a third area command, joining the Coast Guard's Pacific and Atlantic Area Commands.

U.S. Navy Forces. In addition to surface fleet issues, a focal point regarding potential U.S. Navy force projection through the Arctic region is the submarine fleet. Pending a completely ice-free operating environment, submarine force projection would require some degree of ice operations capability. During the Cold War era, the Sturgeon class submarines were designed and built to operate in the Arctic's ice conditions. Today's Ohio Class SSBNs are not designed for ice operations, and there are only a few Los Angeles Class submarines with some modest under-ice capability.

With the end of the Cold War, the Navy's submarine force under-ice skill levels and experience have continued to decline. With a smaller submarine fleet already hard pressed to perform all of the operational tasking needed by the unified commanders and the National Command Authority, fewer and fewer missions have been conducted in the Arctic region.[22] Perhaps a future opportunity for force projection and strategic agility through the Arctic Sea routes will cause a revival in U.S. Navy submarine under-ice operations and capabilities.

U.S. Transportation Command and Arctic Sealift. U.S. national security strategy now rests primarily on projection of personnel and equipment to a theater of operations. That involves moving assets not only out of the United States, but also forward-deployed forces from one overseas location to another.[23]

The requirement of providing initial surge shipping followed by resupply shipping will likely continue to be the sealift plan for crisis response well into the future. Of note is that with the completion of the fleet of large, medium-speed, roll-on, roll-off vessels, the Military Sealift Command's fleet makeup will probably not change much during the first half of this century.

With the potential use of Arctic Sea routes for sealift, TRANSCOM should consider two issues in a future course of action planning process. First, recognizing that the merchant shipping fleet would likely develop ice-capable ships when the Arctic routes connect the Pacific and Atlantic, TRANSCOM should both monitor and encourage industry ship design and building of those ships for future military sealift use. Second, if in fact the Arctic region links the hemispheres of the world, TRANSCOM should then consider the establishment of a major Afloat Prepositioning Force (APF), ready to respond to crises in multiple areas of the world, from a position on top of the world.

Counter Views

The notion of global warming causing the Arctic icecap to melt, and further opening up the Arctic Sea passages, is likely to generate views questioning both the possibility and feasibility of force projection and strategic agility via the Arctic region. Counter views might include the following arguments: (1) The Arctic warming phenomenon is actually a cyclical weather pattern, which causes the Arctic to warm and cool and warm again over hundreds or thousands of years; (2) Force projection won't really be a possibility until the passes are completely ice-free on a year-round basis. There won't be enough ice-capable ships or icebreakers until then, and plans should be based on year-round use, vice based on a few months as a starting point; (3) Force projection through the Arctic in ten to fifty years is well outside budget planning cycles.

These and other arguments are all valid, in varying degrees. Even so, the idea of force projection and strategic agility through the Arctic region would provide new and significant advantages to the combatant commanders in executing U.S. national security strategy. The validity and benefits of building the Panama Canal were questioned, too.

Conclusion

From the scientific standpoint, global warming is and will continue to cause the Arctic ice to melt. From the scientific standpoint, the Arctic Sea routes will eventually be ice-free on a year-round basis.

U.S. interests in the Arctic region will significantly increase during the 21st century. Some estimates indicate that the oil reserves in the Arctic Circle are equal to or even greater than the known reserves of the rest of the world.[24] Oil and other natural and living marine resource interests, along with freedom of navigation interests, will drive increased presence of U.S. joint forces. In essence, the blue water Arctic will be new global space, and the U.S. will be there.

The ability to project forces through the Arctic will likely coincide with increased U.S. presence in the region. For the combatant commanders, force projection through the Arctic will provide flexibility in operational design, and advantages through operational functions. Strategic agility through the Arctic will directly support the national security strategy, in a complex global security environment which will often require U.S. military rapid response, operational reach, and force sustainability.

NOTES

[1] Ross Gelbspan, The Heat is On, (Reading, MA: Perseus Books 1997), 74.

[2] Eugene Linden Churchill, "The Big Meltdown," TIME, (4 September 2000): 54-55.

[3] John Houghton and others, eds., Climate Change 1995, The Science of Climate Change, Summary for Policy Makers, Intergovernmental Panel on Climate Change (Cambridge, MA: Cambridge University Press, 1996), 10.

[4] Donart Pharand, The Law of the Sea of the Arctic, (Ottawa, Canada: University of Ottawa Press, 1973), 27-30.

[5] Ibid., 22-24.

[6] Lara Ellsworth-Jones, "Resistance Melts at Northwest Passage," Daily Express News, 16 June 2000, <http://www.lineone.net/express/00/06/16> [13 April 2001].

[7] Pharand, 51.

[8] Ibid., 51-53.

[9] Ibid., 43.

[10] DeNeen L. Brown, "Piercing the Arctic's Icy Unknown," The Washington Post Online, 20 August, 2000,< http://www.washingtonpost.com/wp-dyn/articles/A36939-2000Aug19.ntml > [28 August 2000].

[11] Pharand, 64.

[12] David L. Larson, Security Issues and the Law of the Sea, (Lanham, MD: University Press of America, 1994), 173-174.

[13] Ibid., 174-175.

[14] The White House, National Military Strategy of the United States of America, (Washington, DC: 1997), 3.

[15] Ibid., 3.

[16] The White House, A National Security Strategy for a Global Age, (Washington, DC: December 2000), 19-20.

[17] Author's estimate of sealift requirement percentage, based on multiple sources. Actual percentage required is probably closer to 95%.

[18] Chet Helms, "Operational Factors," Naval War College Joint Military Operations Handout.

[19] Ibid., 9.

[20] Bryan Knowles, "Was the Transfer of the Panama Canal a Mistake?" SpeakOut.com, November 2000, <http://www.speakout.com/issues/briefs/1100> [9 April 2001].

[21] Jonathan Berkson <Jberkson@comdt.uscg.mil> "Ice Free Arctic Symposium Message," [E-Mail to Dave Hill <anchordetail@earthlink.net> 12 April 2001.

[22] Don Walsh, "A Farewell to Ice," Sea*Power, Navy League of the United States, November 2000/December 2000,
<http://www.navyleague.org/seapower/november2000/December00/walsh.htm> [13 April 2000].

[23] David G. Harris and Richard D. Stewart, "U.S. Surge Sealift Capabilities: A Question of Sufficiency," Parameters, (Spring 1998): 67-83. < http://carlisle-www.army.mil/usaws/Parameters/98spirng/harrishtm> [13 April 2001].

[24] Pharand, xxi.

BIBLIOGRAPHY

Berkson, Jonathan. <JBerkson@comdt.uscg.mil> "Ice Free Arctic Symposium Message." [E-mail to Dave Hill <anchordetail@earthlink.net>] 12 April 2001.

Buderi, Charles L. O. and David D. Caron, eds. Perspectives on U.S. Policy Toward the Law of the Sea: Prelude to the Final Session of the Third U.N. Conference on the Law of the Sea. Honolulu, Hawaii: Law of the Sea Institute. 1985.

Churchill, Eugene Linden. "The Big Meltdown." Time. (4 September 2000): 54-55.

Ellsworth-Jones, Lara. "Resistance Melts at Northwest Passage." Daily Express News. 16 June 2000. Available [Online]: http://www.lineone.net/express/00/06/16 [13 April 2001].

Gelbspan, Ross. The Heat is On. Reading, MA. Perseus Books. 1997.

Harris, David G. and Richard D. Stewart. "US Surge Sealift Capabilities: A Question of Sufficiency." Parameters. US Army War College Quarterly (Spring 1998): 67-83. Available [Online]: http://carlisle-www.army.mil/usaws/Parameters/98spring/harrishtm .

Helms, Chet. "Operational Factors." Unpublished Handout, U.S. Naval War College, Newport, RI.

Houghton, John and others, eds. Climate Change 1995, The Science of Climate Change, Summary for Policy Makers. Intergovernmental Panel on Climate Change. Cambridge, MA. Cambridge University Press. 1996.

Knowles, Bryan. "Was the Transfer of the Panama Canal a Mistake?" SpeakOut.com. November 2000. Available [Online]: http://www.speakout.com/issues/briefs/1100 [9 April 2001].

Larson, David L. Security Issues and the Law of the Sea. Lanham, MD. University Press of America. 1994.

Pharand, Donart. The Law of the Sea of the Arctic. Ottawa, Canada. University of Ottawa Press. 1973.

The White House. A National Security Strategy for a Global Age. Washington, DC: December 2000.

The White House. National Military Strategy of the United States of America. Washington, DC: 1997.

Walsh, Don. "A Farewell to Ice." Sea*Power. (November 2000/December 2000). Available [Online]: http://www.navyleague.org/seapower/november2000/December00/walsh.htm. [13 April 2001].

www.ingramcontent.com/pod-product-compliance
Lightning Source LLC
Chambersburg PA
CBHW081821170526
45167CB00008B/3497